Wood Burning

with
Sue Waters:

RURAL
SCENES

Text written
with and
photography by
Joanne Tobey

Schiffer Publishing Ltd

77 Lower Valley Road, Atglen, PA 19310

Contents

Printed in China.
ISBN: 0-88740-569-X

Published by Schiffer Publishing, Ltd.
77 Lower Valley Road
Atglen, PA 19310
Please write for a free catalog.
This book may be purchased from the publisher.
Please include $2.95 postage.
Try your bookstore first.

We are interested in hearing from authors
with book ideas on related subjects.

Introduction

If you've been introduced to the art of wood burning, then you know what an enjoyable—and addictive—art form it can be. One of the pleasures I take in wood burning is that it always teaches me something new. Every project is a fresh start, with opportunities to "draw" new objects and textures.

If you've been looking for new ideas to stimulate your wood burning work, this book is for you. In it I tackle one of my favorite subjects: barns in their country settings. Work through the step-by-step exercises in these pages with me, and you'll learn how to produce realistic wood siding, stone walls, meadows, and mountains—everything you need to put together a handsome outdoor scene.

I've included patterns for five wood burnings. When you've finished with them, you'll be ready to create your own works of art.

Have fun!

Supplies

Wood burning tool (preferably a solid state unit, with variable degree settings)

Wood (bass-wood plaques or bark-on planks, available at craft stores or from cut lumber)

Sandpaper or sanding disc

Tracing paper

Graphite paper

Tape

Soft-lead pencil

Metal-edged ruler (optional)

Varnish (non-yellowing, brush-on. I prefer Satin Finish Right Step, available at craft stores.)

Tools

There are two different types of wood burning tools: the single hand-held unit that most of us are familiar with (it looks like a large plastic pen with an electrical cord at one end and a brass burning tip at the other), and the solid-state machine with temperature control and changeable hand and tip pieces (the "tip" is what you actually use for "drawing"). When I first began to wood burn, I used a single-unit tool; I advanced to the solid state when I wanted to work more quickly. It's the type of unit you'll probably want to use for the wood burnings in this book. I've provided you with a list of supplies. (The single-unit tool is adequate for burning; you'll simply have to work a little bit more slowly and cautiously to have the type of control necessary for fine details.)

I usually set the temperature on the solid-state unit somewhere in the middle, varying it occasionally, depending on the technique that I'm using and whether or not I want a dark or a light burn. (High heat produces a dark burn quickly; low heat produces a light burn.) I always use a low heat setting when I'm doing small, intricate work, so that I have better control over what's happening on the board.

There are many tips that can be used with a solid-state unit. In this book I'll be using three: the knife tip, the shading tip, and the writing tip. The knife tip has a sharp straight edge; it actually puts a bit of a depression into the wood, and therefore is most useful for straight, dark lines. The writing tip is a thick, curved wire, with a surface area similar to that of a pen; it burns a fine line. The shading tip has a "spoon" on the end of it, a broad surface that allows you to burn a fairly wide strip.

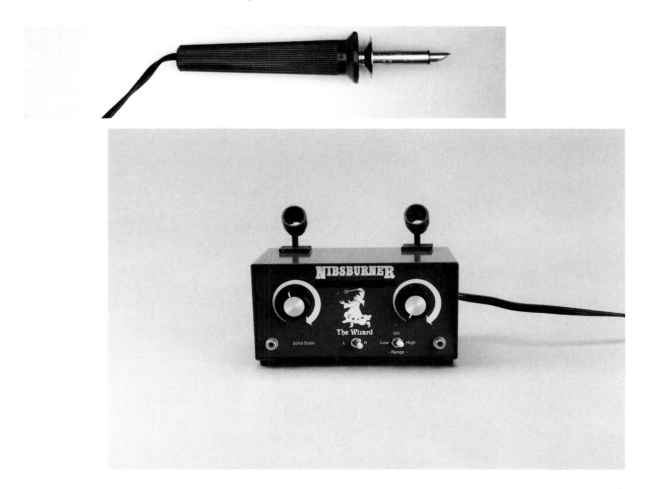

Wood

There are two types of wood that I favor for flat wood burning: birch plywood and bass wood. Both are light-colored woods, with smooth grain that's easy to burn over. (If you try to burn on a rough-grained wood, such as oak, you'll find that your burning tip catches, producing a very uneven line. Pine has a pitch that makes burning difficult, and some woods are so dark that the burn does not show well). Birch plywood is readily available at most lumber yards; it's easy to cut the board into picture sizes, using a hand saw or circular saw. Bass wood usually comes in planks up to fourteen inches wide and three-quarters of an inch thick. This wood is great for making your own plaques. You can cut the board to size and rout the edges to give them detailing.

I frame a lot of my work, so it's important that I cut it into standard-size pieces before I get started. Custom frames for odd-sized pieces of wood are expensive. Below is a list of standard picture sizes. These are the same sizes used for framing artist's canvas.

5 x 7	11 x 14	12 x 24
8 x 10	12 x 16	18 x 24
9 x 12	16 x 20	20 x 24

Troubleshooting tip: When measuring a frame for your work, measure the *inside* dimensions of the opening. There is usually a one-quarter inch easement for the work to be set in.

Preparation of the Wood

The best surface to burn is a smooth one. A well-sanded board allows the burning tip to glide smoothly, producing even, controlled lines. Before you sketch out the piece, go over the board with a fine-grit sandpaper (coarser grit if the board is very rough). I use sanding discs, which I buy at a craft supply store; a regular sanding block or palm sander will do just as well.

Our first project together will introduce you to wood grain: how to produce it and how to make your wood-sided buildings look as natural as possible. We'll also tackle various textures, such as leaves, pottery, glass, and grass. With this first wood burning under your belt, you'll find the projects that follow challenging but fun.

Enjoy!

Patterns

Pattern enlarged 134%.
Reduce 75% for original size.

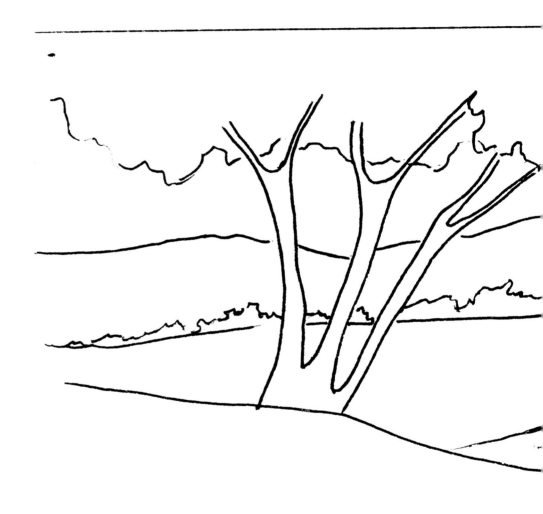

Pattern enlarged 126%.
Reduce 79% for original size.

Barn wall with flower pots, etc.

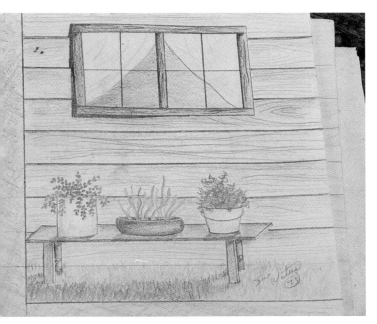

Here's the drawing on which I've based the pattern for our first project together. You might eventually want to produce a drawing like this for one of your own creations; I sometimes find that making a complete picture helps to add detail to the wood burning. A simple pattern has its advantages, though: Tracing only the essentials cuts back on the amount of time you have to put into preliminaries, and it gives you the freedom to add your own personal touch to your work as you burn.

The first step in transferring the pattern to the board is to trace it. I use a number-two pencil and a sheet of tracing paper laid down over the pattern.

Next, I tape the tracing paper (with pattern) to the top of my board. Taping only one edge holds the paper steady while allowing me to flip it up and check to see the progress on my tracing.

Here's the shading tip that I'll be using first on this burning.

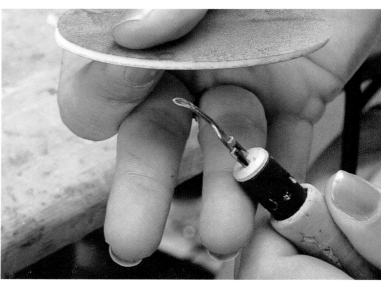

With the patterned tracing paper taped down, I slip a sheet of graphite paper underneath and start transferring the pattern to the board. I like to use a colored pencil or firm-felt-tip marker for this tracing, so that I can see more clearly what I've done.

If you are using a single-unit tip for the first time, you'll want to sand the sharp edges, so that they don't gouge the wood. You should also lightly sand your tip if it starts to drag while you're burning. That drag is the result of carbon build-up, and it will slow your work considerably.

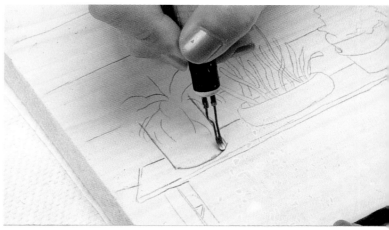

This is what the traced pattern looks like on the board: far less detailed than the original drawing, but the essentials are there.

Unlike in painting, where you start in the background and build to the foreground and highlights, wood burning requires that you start in the foreground and work back to prevent overlap. I begin with the crock of flowers on the left side of the bench.

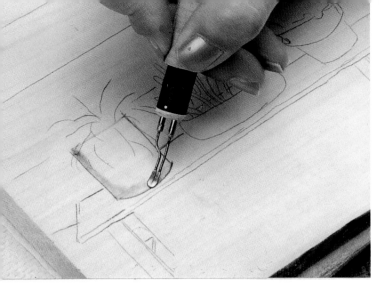

This is an old, manila-colored crock, so I shade it lightly, moving my burning tip quickly to prevent too dark a burn.

I burn in the plant one stem at a time, adding small ovals for the leaves as I go.

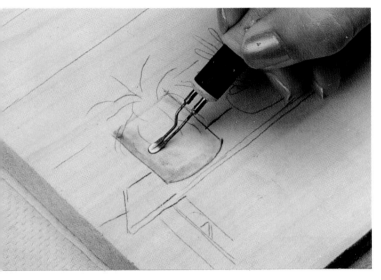

I've decided that my light source is from the right; that means that all objects, including this crock, should be darker on the left-hand side.

By staggering the placement of the leaves on the stems and filling some in while leaving others "empty," I produce a natural-looking plant with lights and darks that add interest.

By burning the center of this crock only lightly and making the edges darker, I've achieved a nice round look.

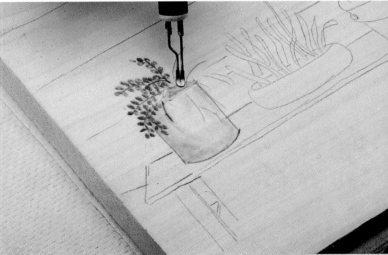

The way the leaves overlap each other and the stems drape the pot gives the plant fullness.

19

In this close-up, you can see the variation in the shades and shapes of the leaves.

The right edge of the bowl shows a burnover: the heat of the tip has made a fuzzy mark.

The next bowl is clay and should be darker than the pot. I draw it in with the shading tip, leaving an unburned band around the top as a line of reflection.

A quick sanding gets rid of the problem.

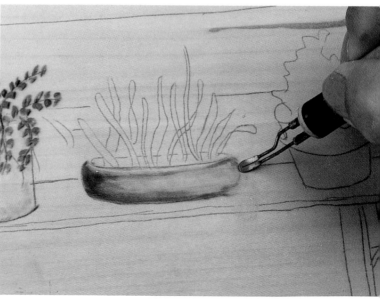

Like the pot, the bowl is darker to the left, with the shading becoming lighter toward the center.

I redefine the edge with a slow, controlled stroke of the shading tip.

Shading the back edge of the bowl—visible between the stems of the plant—not only reflects a realistic play of light; it helps set off the plant.

The shading in this plant is soft, done with light strokes, and that helps distinguish it from the preceding plant.

I run a dark line to define the bowl edge below the plant stems.

This metal bucket will be a shade between the pot and the bowl. I start by outlining it and its handle.

Each stem is one smooth, upward stroke. Applying a little more pressure on the end of the tip helps it to darken the left side of the stem.

A metal bucket has a rolled lip; I define it by burning two curved lines, parallel to each other, with a band of unburned wood in between.

21

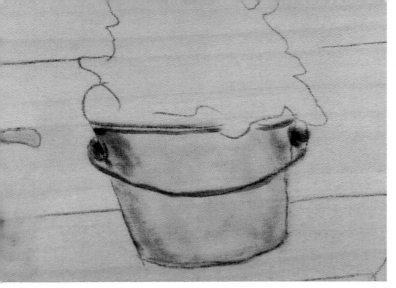

Here's the bucket shaded dark on the left to give it roundness.

Although it's not on my drawing, a "dent" would add character to the bucket, so I put one in with a single, slow stroke.

A lighter stroke to the edge of the "dent" gives it depth.

I switch to the writing tip for the leaves of this plant. I want it to be a full, leafy plant, so I "scribble" in the foliage, using a random, up-and-down then side-to-side motion.

Once the leaves are roughed in, I go back to the shading tip to add dark areas. Contrast—*light* lights and *dark* darks—is what makes a wood burning look good.

The three plants and their pots show the variety of textures and values you can obtain with just two wood burning tips.

I've switched to the knife point to lay in the horizontal lines that make up the bench and wood siding.

Take your time burning in the segments of the board lines that go behind the plants.

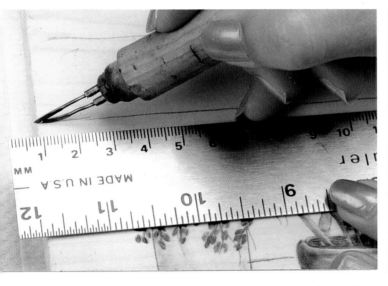

Along with using the burning tip against a ruler, you can produce straight lines by resting your pinkie against the ruler (useful if your ruler isn't metal).

The knife tip actually cuts into the wood; once you have the cut started, it's easy to continue it . You may find that sometimes the tip will skip a little or produce a different burn because of the grain of the wood. This simply adds character to the wood burning.

Since the wood is weathered and therefore uneven, a bit of unevenness in your lines isn't anything to worry about—it's good!

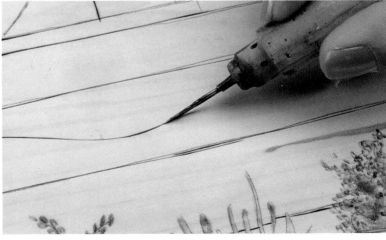

Look at an old board for reference if it helps you to burn in the grain of the siding. I find, however, that one of the fun things about burning wood grain is that you can create your own patterns.

I burn a wavy line down the length of one "board."

Then go back outside your first line and trace the pattern bigger, to fill in the "board."

Then I go back, and with my burning tip inside the first line, repeat the pattern.

When I reach the edge of a "board," I simply end the line and pick up on the other side.

Build on the pattern, tracing lines to the inside of each previous line, until you create an "eye."

Here's the finished "board": the wavy lines and slight variations in the thickness make it look just like a real grain pattern.

Fill in the rest of the "boards" the same way. When you get down around the plants . . .

I've switched to the writing tip to put in the grass below the bench. Grass is done with an up-and-down "scribbling" motion, and I want it burned in before I try to put in the grain of the "board" behind it.

. . . skip over the leaves . . .

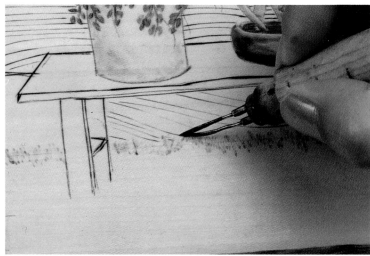

With the grass in, I use the knife point to lightly burn in the grain on the bottom "board." Because it's behind the bench, and because I don't want it to detract from the interest of the plants, I limit the "board's" grain to a simple series of diagonal lines.

. . . and resume the pattern on the other side.

Here's how the board looks with the plants complete. The wood grain is burned in and the bench is outlined. I've used the shading tip to draw a dark stroke under each "board," representing the shadows produced by overlapping boards.

I use the knife tip to burn in the wood grain of the window frame. This grain is done just like that of the wood siding except the lines are vertical instead of horizontal.

Because my light source is from the right, even the window frames are darker along the left edges.

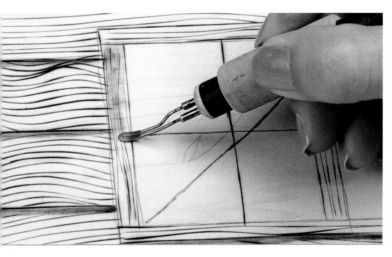

I use the shading tip to add values to the window frame, too, helping it to stand out.

I also darken them where two "boards" meet: for example, in the upper left-hand corner.

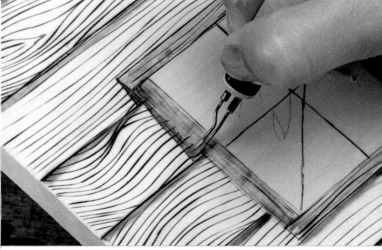

Move the tip slowly enough that you get a good, dark, defining burn. If the burn isn't dark enough to make the window frame seem to pop out, go back over it again.

The shadow that the window casing drops along the barn wall follows the contours of the wall. Each clapboard is thicker and raised at its bottom edge, so the shadow is triangular, rather than straight.

I burn in a narrow triangle on each "board."

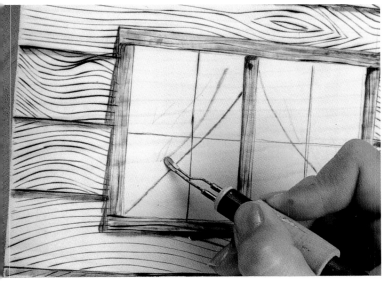

Next, I shade the window curtains, burning in some curved lines to suggest draping.

Now I burn in the window glass. I want the shading very light, so I move my burner quickly, in parallel diagonal strokes.

To give interest to the picture, I burn the glass on the other side of the munton with diagonal strokes angled in the opposite direction. (The natural grain of the board adds interest, too: you can see here how it produces a wavy pattern in the lower right-hand corner of the left-hand window.)

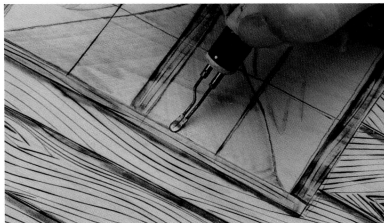

I burn a dark line at the bottom of the glass to define the sill.

With the window complete, I've decided to add more shading to the barn siding. Using the shading tip, I give the siding color with a few broad strokes.

27

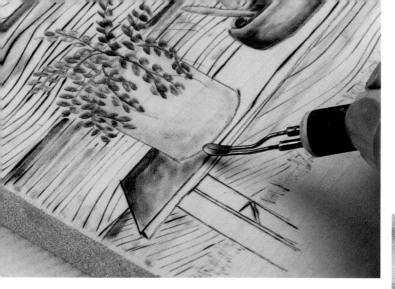

I shade the bench so that the top surface is dark and really sets off the plant containers.

The shading deepens to the left of the first crock and mirrors its shape.

I work my way down the bench, burning in shadows for each of the plant containers. On the bench legs, I work carefully, touching the tip of my burner to shade the left edges of the wood.

You can see where I've left the front edge of this cross-piece light, because the sun is hitting it.

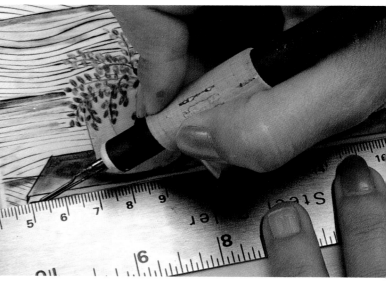

I use my ruler and the writing tip to darken the bottom edge of the bench top.

With the writing tip, I make some upward strokes to bring grass around the legs of the bench.

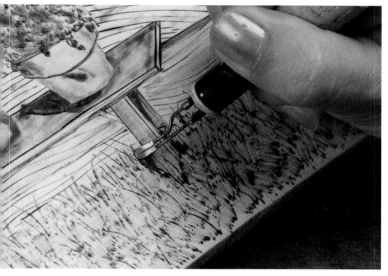

The spoon tip allows me to lay in the shadow the bench legs produce against the grass.

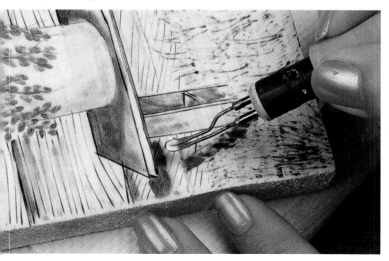

The shadow that the bench throws against the wall reflects the shape of the bench: a narrow strip running on a diagonal across the grass and (roughly) vertically up the wall reflects the legs; a horizontal strip against the wall reflects the bench top.

The shading tip also allows me to add some texture to the grass.

The variety of textures and patterns makes this a very pleasing wood burning. All it needs now are two to three coats of matte varnish and a frame.

Wall with stones

This project, with a panoramic view of mountains, fields, and an old barn full of hay, introduces you to a new texture: rock walls.

I've traced my pattern onto the board and have blocked in a lot of the background: the mountains, done with the shading tip; the trees, done with the writing tip and darkened with the shading tip; and the barn, outlined with the writing tip. I've started the wall in front, drawing the top line of stones with the shading tip.

Variety in the shapes is the key to a life-like stone wall. You can see here how I've burned in small and large rocks, and that the outlines aren't perfectly oval—they have a certain wobbliness that imitates the natural unevenness of stone.

Where the rock wall is broken, I draw some curved lines to represent loose rocks that have tumbled into the grass.

I vary the width of my lines, too. A heavier stroke on the underside helps round out this rock.

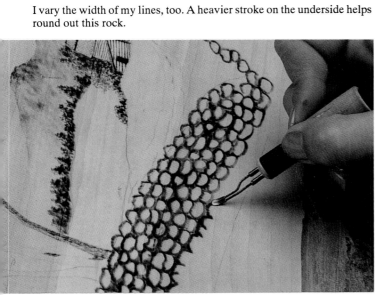

I build the wall from the top down, leaving the last line of stones open at the bottom, so that I have a clear space into which I can pull up some lines for grass.

With the stones burned in, I go back , adding some light shading toward the bottom of each rock. I move the tip rapidly, so that I simply "smudge" the rock.

31

The shading adds a lot of texture. Compare the left, shaded side of the wall with the unshaded side to the right of the break.

With the shading tip, I darken the right-hand sides of the tree branches, using smooth, long strokes.

Now I switch to the writing point and burn in the large tree in the foreground. I use short, vertical strokes to suggest bark.

Because I want a loose, full look to the foliage on this tree, I "scribble" it in with the shading tip.

Here's a close-up of the tree, showing how the short strokes give it texture.

Darker foliage helps a branch to show clearly, giving the wood burning some good contrast.

The grain of the wood produces a slight color change at my horizon line. That works to my advantage, adding a purple hue to the mountains.

I drawn a line underneath the edge of the barn's roof to suggest shadow. Because this is detailed work, I keep my burner on low or medium heat to prevent burnover.

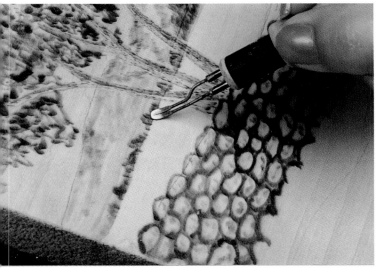

With the mountains burned in, I use short, dark strokes to suggest a tree line.

A dark burn along the line of the shed roof helps to define it.

A wood burning isn't complete until you want it to be. Here I've decided that the wall needs more texture, so I've gone back to darken the spaces between the stones, to help round them out.

I add a few light strokes to the roof.

I burn another line of dark, vertical strokes to suggest the grass on the left-hand side of the road, and then work to fill in the grass of the field. As in painting, objects in the background appear lighter than objects in the foreground, so the grass in my far field is very light—I've only burned in a few strokes here and there—and the grass around the road is darker.

There's an overgrown tractor road leading to the barn. I define the center with a broken line of short, vertical strokes that suggest grass.

When I get to the right-hand side of the road, I burn in another curved, broken line to represent grass. Then I continue shading the field, moving the burner in a fairly open, random pattern.

To give the impression that the road is slightly sunken (as it would be from travel), I add some broad, slightly curved strokes that become wider as they join with the line of grass.

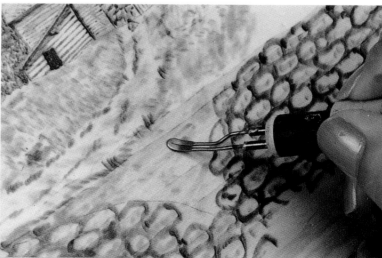

To make the grass just behind the wall seem closer than that around the barn, I make the strokes wider and longer.

The grass in the immediate foreground, in front of the rock wall, is the darkest and most well-defined. I use vertical strokes to work it up into the last line of stones . . .

. . . and continue to fill in the foreground, working right to the edge of the board.

The finished product. The variation in the values and textures gives this wood burning real depth and pulls you into the picture.

Barn

Here's what we're aiming for in this project, an autumn scene of an old barn set among trees. Burning this picture will develop your sense of how shadows fall and will introduce you to producing textures that look like metal.

After the pattern is traced, I go to work with my straight edge and writing tip, burning in the lines of the barn boards.

With the shading tip, I darken the openings, the windows, the door . . .

. . . and the underside of the roof projection. Then I begin to add values to the siding, shading the boards horizontally. I darken one board at a time.

The unevenness in tone that comes from varying the speed at which I move the burner gives the wood a great natural look.

Where the barn boards run vertically, such as on the loft door, I switch the direction of my stroke.

Broad dark lines that follow the direction of the roof shade one eave. I add a long dark line to further define it.

The light in this picture is falling from the right, so I draw the shadow that the eave on that side of the barn produces.

Where the door falls against the wall of the barn, it produces a shadow. Notice also in this picture a dark horizontal band in the door wood, a little bit more than halfway down. This is the result not of a burning technique but simply the variation in the grain of the wood and how it picks up the burn. There's nothing wrong with such "imperfections." They add character to the piece.

I work around to the side of the building. Because it's in the light, I don't make it quite as dark as the front of the barn.

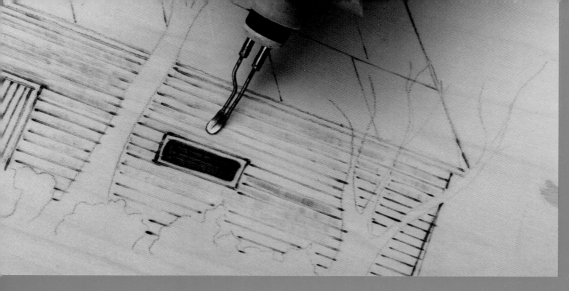

Once again, I shade the siding board-by-board, using horizontal strokes.

A shadow to the left of the side door cleans the edges up and helps to give the building depth.

I put a shadow line under the eave. It is a rather heavy shadow, as the sun is shining on this face of the barn.

Next, I burn in the roof. It's made of corrugated metal sheets. I've outlined the individual roofing sheets, and now I'm burning in the parallel lines that represent the corrugation. I keep the roof very light—because the sun is shining on it, and because I'm going to be placing shadows from the trees here and want to produce a nice contrast between the two.

"Ruffles" at the bottom of the metal sheets simulate the corrugations.

The grain of the wood helps add variety to the values in the roof.

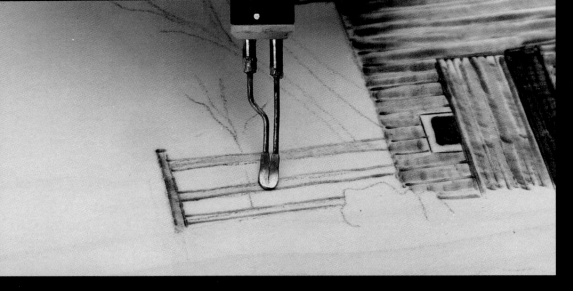

While I have the shading tip on my burner, I burn in the section of fence alongside the barn. Rails are basically a horizontal stroke or two; the fence post is a vertical stroke.

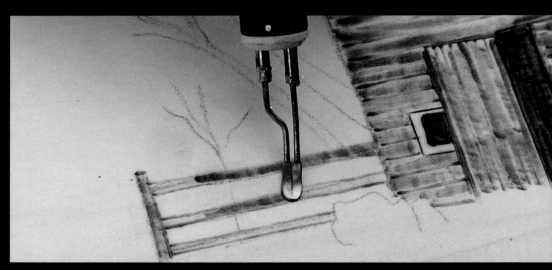

I darken the rails for about 3/4 of their length, where the shadow from the barn falls on them.

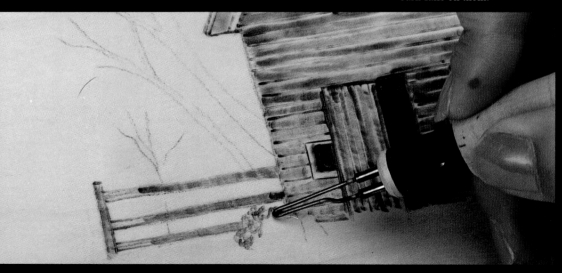

I switch to the writing tip and "scribble" in the bushes. The bushes on the shadow side of the barn should be darker than those on the long side; to darken them, I simply move the burner more slowly.

Making the bushes along the front side of the barn even darker than the wall gives the wood burning necessary contrast. The bushes along the side wall are in the light, and therefore need more detail: a greater difference between the darkest areas and the highlights. The darkest shading should be along the undersides of the bushes, where the foliage meets the ground.

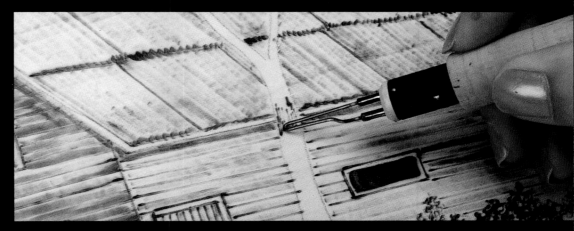

An up-and-down motion produces a bark-like texture on the tree.

To make the tree appear round, I keep the right edge lighter than the left and darken the base of the trunk, where the brush casts its shadow.

Burning in the small branches on the trees is fun. They're a lot of random curved lines.

The shading tip helps me to add the finishing touches to the shadows on this tree.

With every object I burn in, I keep the shadows to the left side.

With the trees complete I burn in their shadows against the barn. The angles of the shadows should reflect the angles of the objects they're falling on. Here the shadow falls vertically on the vertical wall of the barn

. . . and falls diagonally across the roof.

I burn in shadows reflecting the major divisions of the tree branches . . .

. . . and suggest the shadows for the smaller branches.

The shadows help set the trees off from the barn.

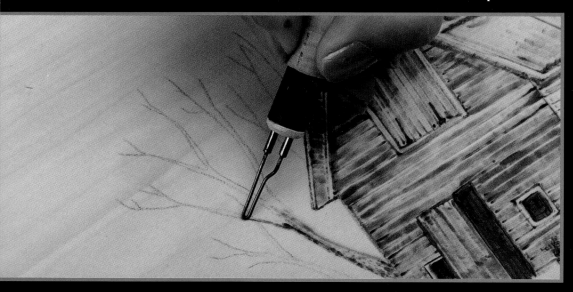

I use the writing tip to burn in the trees along the front of the barn. Here the challenge will be to make the many branches overlap in a realistic way.

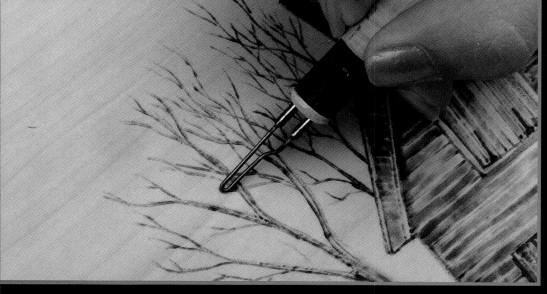

I've burned in the main trunks of two trees and am now adding the finer branches. I burn this branch, which belongs to the tree in the background, behind the other tree, breaking the line when I come to one of its branches and then resuming the burn on the other side.

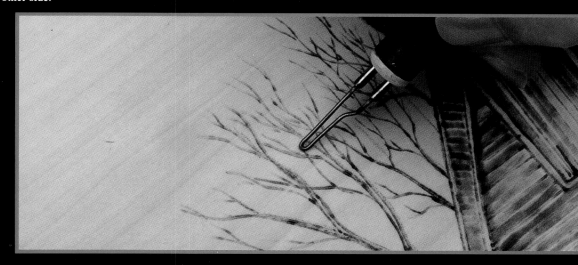

This branch, which belongs to the tree in front, is burned over the branches behind it.

With the background items finished, I lay in the grass. Because this is an autumn scene, I want to suggest that leaves are covering the ground. A very loose, random pattern helps to suggest that.

A close-up of the squiggles that compose the grass.

With the line work done, I go back in with the shading tip and add the shadow that falls from the front side of the barn.

I let the grass pattern "unravel" toward the front of the picture, so that it appears to fade off into the edges of the board.

Miscellanous Barn

Here's a barn scene that incorporates some of my favorite elements: mountains, fields, and an old tree. What's new is the way I'll be burning the grass: I'm using a fine-art style, rather than a literal technique. I'll suggest the grass with large areas of shading, defining blades only in certain areas.

I've outlined the barn with the writing tip, burned in a few birds, and have used the shading tip to produce mountains and the trees in the background. The light source is from the right.

Here's a close up of the tree, showing how a random pattern of squiggly lines, along with a variety of tones (lights and darks) helps to give it fullness.

I'm laying in the field. It's basically a network of openings outlined with light, jagged shading, topped with dark vertical lines that represent clumps of grass.

Breaking the lines of grass suggest that the road is overgrown.

I use the edge of the shading tip to burn in the grass lines.

I work the grass right down to the edge of the board, slowing my burning to produce darkness on the undersides of some of the grass clumps to "ground" them. The road widens as it comes forward.

The grass along the farm road is done the same way, using the shading tip, with short, vertical strokes.

Now I burn in the two dirt tracks of the road. Since this road is well-worn, the tracks are sunken, and I show this by curving the burn lines.

A few dark specks on the road suggest rocks.

I define the eaves by burning in a dark shadow line.

Now I go back and shade the other track.

A dark, uneven line defines the wavy edge of the corrugated metal roof.

The barn is the focal point of the piece, which its placement on top of a hill accentuates. I begin to shade it, giving it depth and texture. I follow the outline of the boards with my shading tip. The grain of the wood adds nuance to the texture.

Long, parallel strokes define the metal sheets that compose the roof.

Although I burn some dark areas into the roof, I keep it lighter than the walls of the barn to distinguish the two.

A dark line separates the roof's edge from the gable trim.

Although the barn is well-shaded, it seems to lack contrast. I'll go back in with the shading tip to darken and add shadowed areas

The door has high contrast, with the cross parts of the x's standing in sharp relief against the vertical boards to the back. Accentuating the contrast helps your eye pick out these features and makes the door look life-like.

I've added some shading to the grass on the left of the barn, and that helps balance the darkness of the mountains.

The right edge of the barn and the mountains seem too much the same color. I can darken the mountains or the barn, and I decide to darken the mountains in order to make the barn pop out.

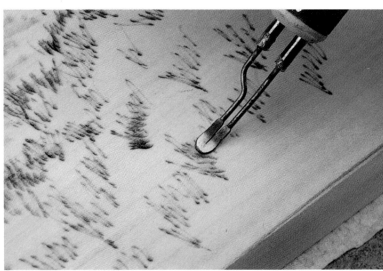

I repeat my grass burning procedure for the right-hand side of the board, scribbling in vertical strokes to represent clumps of grass . . .

To balance the picture, I darken the mountains on the other side of the barn, too. The picture would look awkward if I had darkness only on the mountains on the right-hand side.

. . . and adding a loose pattern of shading for the background.

Using the shading tip, I darken the edge of the road, to separate it and the grass.

I use the writing tip to add a few well-defined lines of grass.

The finished picture. I'll frame this, using finish nails to tack the frame onto the varnished board.

Squirrel on fence Post

Burning this pattern—a squirrel on a fence post—will give you the opportunity to use your knowledge of wood graining and will challenge you to produce the fine detailing of a small animal.

Outline the fence post using the writing tip.

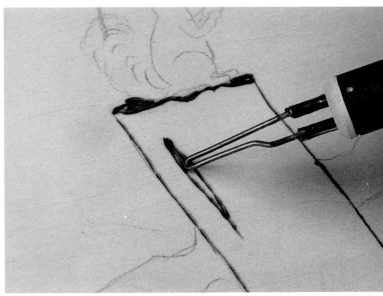

Here's the split nearly complete.

I've darkened the top of the post around the squirrel and am burning in a split in the wood.

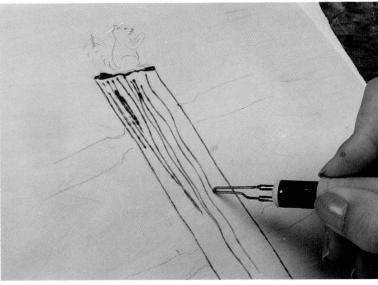

Next I burn in wavy lines to simulate wood grain. This gives the post character.

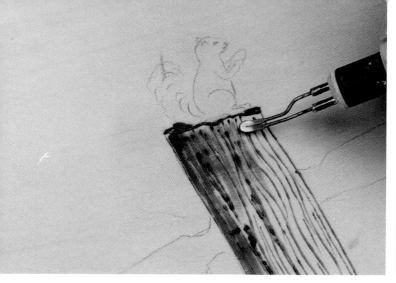

I switch over to the shading tip, working at low heat. The shadow side of the post is to the left.

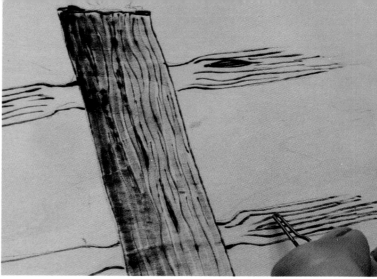

I burn in the grain, placing splits randomly along the rails.

I lighten the shading toward the center of the post by moving the burner faster.

Then I switch to the shading tip to darken the ends of the rails where they enter the post and their undersides.

I'm going to burn the fence rails much the same way, starting with the writing tip to produce a fairly dark outline.

Because the shadows fall to the left, the rails to the left of the post are darker than those to the right.

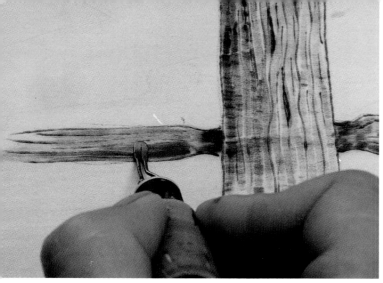

I darken some areas to bring up contrast and help round the rails.

I carry the dark line down around the leg to the foot.

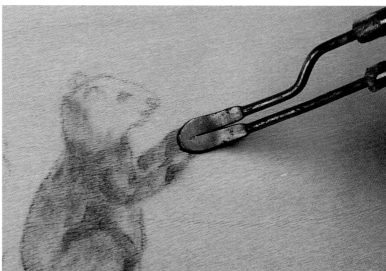

Burning in the squirrel with the shading tip gives it a natural softness. I start by roughing in its coat with short, light strokes.

I fill in the rest of the body hair with short strokes. Letting the burner tip rest on the wood for a second places a "nut" between the squirrel's forepaws.

A dark line above the point of its haunch sets the hind leg away from the body.

A squirrel's tail is a bushy thing, and I try to get that across by burning it in short, curled strokes, first up the front of the tail . . .

. . . then up the back.

I use the writing tip to add a few well-defined tail hairs.

I define the tail by burning in a curved "tail bone."

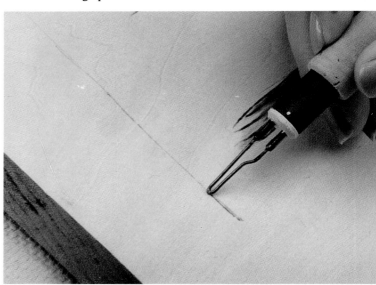

Then I move on to the plant growing alongside the fence. I begin by stroking up the main stalk.

Here's what the squirrel looks like close up, with short, dark, curved lines to indicate ears and a dark spot for his eye and nose.

Clumps of dark, squiggly lines suggest leaves.

I complete one stalk, making sure that it ends above the squirrel, not level with him. If everything in the picture lined up, it would be boring.

And I add a few tendrils around the bottom of the fence post. The one to the right of the posts helps to balance the plant grouping on the left.

It's better to draw in a line too light and go back over it than to make it too dark and wide with the first stroke.

To help ground the post, I shade around the base very lightly.

I fill in the bush, making some branches darker than the others to add interest.

Bringing the shading up around the bush helps to set it into the ground, too.

59

The finished picture is lively, with good contrast.

Gallery